Rahma SAID

AKAP lbc genotypes and implications for GVH research

Rahma SAID

AKAP lbc genotypes and implications for GVH research

AKAP lbc Genotypes and Impact on GVH Research in Hematopoietic Stem Cell Transplant Recipients

ScienciaScripts

Imprint
Any brand names and product names mentioned in this book are subject to trademark, brand or patent protection and are trademarks or registered trademarks of their respective holders. The use of brand names, product names, common names, trade names, product descriptions etc. even without a particular marking in this work is in no way to be construed to mean that such names may be regarded as unrestricted in respect of trademark and brand protection legislation and could thus be used by anyone.

Cover image: www.ingimage.com

This book is a translation from the original published under ISBN 978-620-6-70214-6.

Publisher:
Sciencia Scripts
is a trademark of
Dodo Books Indian Ocean Ltd. and OmniScriptum S.R.L publishing group

120 High Road, East Finchley, London, N2 9ED, United Kingdom
Str. Armeneasca 28/1, office 1, Chisinau MD-2012, Republic of Moldova, Europe
Printed at: see last page
ISBN: 978-620-7-65673-8

Table of contents

INTRODUCTIO N

Minor histocompatibility antigens (AgMHs) are endogenous, polymorphic and immunogenic peptides that can induce cell- and humoral-mediated immune reactions between geno-identical HLA individuals. In hematopoietic stem cell (HSC) allogeneic transplantation, these antigens represent the major target of graft-versus-host disease (GVHD) and graft-versus-leukemia and tumor cell disease (GVL and GVT). The discovery of these AgMHs was made in 1948 by Snell following the triggering of immune reactions and the appearance of post-transplant complications in HSC recipients.

In the present study, we were interested in investigating the polymorphism of the *AKAP lbc gene* encoding the HA-3 minor histocompatibility antigen.

In the laboratory of the Centre National de Transfusion Sanguine (CNTS), we were asked to develop a simple, rapid molecular genotyping technique applicable to this gene to study its molecular polymorphism in the Tunisian population.

The results of this study will serve as the basis for further clinical studies in Tunisian transplant patients.

I.Histocompatibility
I.1.Definition

Histocompatibility or tissue compatibility is the ability of tissues to tolerate each other. It mainly involves two groups of antigens on which the success of a hematopoietic stem cell (HSC) or solid organ transplant depends. The first is the group of major histocompatibility antigens known in humans as HLA (Human Leukocyte Antigen). The second is the group of non-HLA antigens or minor histocompatibility antigens (AgMHs).

I.2.Historical background

The first human bone marrow transplants were carried out in 1957 by Donnall Thomas in New

York, who was awarded the Nobel Prize for Medicine in 1990, resulting in the death of 6 recipients in less than three months (Thomas. ED et al., 1957). They were performed at a time when the notion of histocompatibility did not exist. The foundations of histocompatibility were laid in 1958 by the French physician Jean Dausset (winner of the Nobel Prize for Medicine in 1980) (Dausset.J ,1958).

I.3 Histocompatibility antigens

I.3.1. Major histocompatibility antigens (HLA)

a.General information

Major histocompatibility consists of a group of tissue antigens forming what is known as the MHC or HLA major histocompatibility complex (Cesbron-Gautier.A et al., 2007). Historically, the concept of major histocompatibility was developed over 60 years ago. The leukoagglutination reactions observed with sera from polytransfused subjects and certain sera from multiparous women enabled Dausset in 1954 to put forward the hypothesis of the existence of leukocyte alloantigens, and subsequently to call them HLA antigens (Dausset.J, 1958). The role of the HLA system in the allogeneic response led to the development of tools for analyzing its polymorphism. The introduction of complement-dependent microlymphocytotoxicity made it possible to test the reactivity of immune serum to lymphocytes from unrelated subjects, and thus to describe HLA class I molecules. Activation reactions following mixed lymphocyte cultures

molecules (Terasaki.Pi et al., 1964). Fundamental and clinical studies carried out over the last two decades have identified the main features of this system: codominance alleles, close linkage between the various genes in the HLA region, haplotypic transmission and linkage disequilibrium at certain loci (Colombani.J, 1993; Schwartz.B, 1994).

b.Genetics of the HLA system

With the advancement and improvement of certain molecular genetic techniques, it has become very easy to pinpoint the precise location, arrangement and orientation of HLA genes. These techniques have made it possible to identify numerous genes coding for the classic HLA antigens involved in immune reactions (HLA class I and HLA class II antigens), genes more or less related to the classic HLA genes, and genes apparently unrelated to immune functions but belonging to this chromosomal region (HLA class III) (Schwartz.B, 1994). The HLA locus is located on the short arm of chromosome 6 (6p21.3) and occupies around 5000 Kb (Klein.J and Sato.A, 2000) (Figure 1). The complete sequence of this locus, which has been widely identified, comprises over 200 genes grouped into three classes: class I comprises the HLA genes -A,-B, -C,-E,-F, and -G, class II comprises the DR, DQ and DP genes, and class III comprises the genes lying between the two preceding classes.

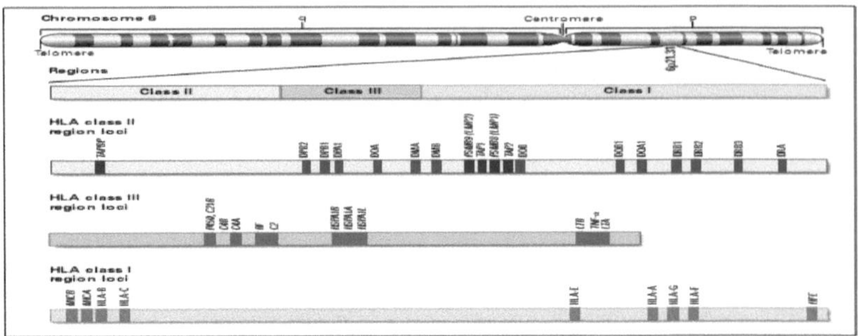

Figure 1: Chromosomal localization of HLA system genes. (Klein.J and Sato.A, 2000)
c.Molecular structure of HLA class I and II proteins

Given their involvement in the majority of immune reactions, HLA class I and II molecules constitute the most extensively studied immunogenetic system to date. They ensure the presentation of endogenous and exogenous peptides to T lymphocytes, with the aim of maintaining the organism's integrity. These molecules also play a crucial role in determining the success or failure of any organ or tissue transplant, since they are the physical carrier of HLA antigens.

❖ **Molecular structure of HLA class I proteins**

HLA class I molecules are transmembrane glycoprotein heterodimers composed of a glycosylated α-heavy chain non-covalently associated with a β2-microglobulin light chain (Klein.J and Sato.A, 2000; Cesbron-Gautier.A et al., 2007). The latter is encoded by a gene located outside the HLA region on chromosome 2. The α chain is made up of 345 amino acids and organized into 3 regions as illustrated in Figure 2 below:

➢ An intracytoplasmic C-terminal region (30 to 35 amino acids).

> A transmembrane region (26 amino acids).

> An extracellular region subdivided into three domains α1, α2
and α3, each of which is around 90 amino acids long. The α1
and α2 domains each contain a di-sulfide bridge.

The three-dimensional structure of HLA class I molecules has been described since 1987. At the top of the molecule is a groove formed by two parallel α-helices, which form the border, and 8 antiparallel β-sheets, which form the floor.

This groove is formed by the α1 and α2 domains and constitutes the peptide binding site (Schwartz.B, 1994).

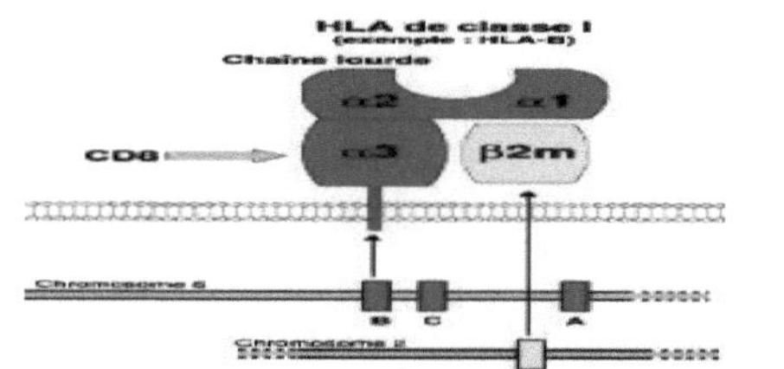

Figure 2: Schematic structure of the HLA class I molecule

(Klein.J and Sato.A, 2000 ; Cesbron-Gautier.A et al ., 2007)

❖ **Molecular structure of HLA class II proteins**

HLA class II molecules are formed by two non-covalently associated transmembrane glycoprotein chains (Figure 3) (Klein.J and Sato.A, 2000 ; Cesbron-Gautier.A et al ., 2007). Each HLA class II molecule consists of an α chain and a β chain, each of which comprises :

> Two N-terminal extracellular domains α1 and α2 for the α chain and β1
and β2 for the β chain. Each domain is about 90 amino acids long.

> A transmembrane domain.

> A C-terminal intracytoplasmic domain.

The HLA class II molecule features symmetry of the α1/β1 domains on one side and α2/β2 on the other. Its three-dimensional structure is comparable to that of the HLA class I molecule. the juxta-membrane α2 and β2 immunoglobulin domains support the N-terminal α1 and β1 domains, which together with a folded β platform and two α helices form the peptide presentation site (Klein.J and Sato.A , 2000 ; Cesbron-Gautier.A et al., 2007).

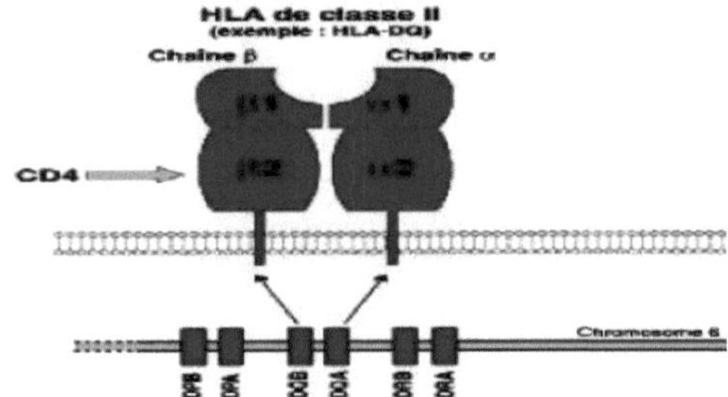

Figure 3: Schematic structure of the HLA class II molecule (Klein.J and Sato.A, 2000 ; Cesbron-Gautier.A et al., 2007)

d.Tissue distribution of HLA class I and II molecules

❖ **Tissue distribution of HLA class I molecules**

Classical HLA class I molecules are expressed on the surface of all nucleated cells with the exception of the central nervous system (Colombani.J, 1993). Quantitative differences are observed according to cell type and locus. In contrast, expression of HLA-G, HLA-E and HLA-F is restricted to certain tissues (Carosella.ED et al., 2000).

❖ **Tissue distribution of HLA class II molecules**

HLA class II molecules are expressed more restrictively than HLA class I molecules. Only antigen-presenting cells express these molecules, namely B lymphocytes, macrophages, langerhans cells, dendritic cells and activated T lymphocytes (Colombani.J, 1993).

I.3.2. Minor histocompatibility antigens or non-HLA antigens

The notion of non-HLA antigens was particularly developed to designate the group of antigens encoded by genes inherited independently of the classical HLA region. This group, later renamed the minor histocompatibility antigen group, was discovered following the occurrence of severe complications in patients transplanted with hematopoietic stem cells (HSCs) from geno-identical HLA donors. In what follows, we present the genetic and biochemical characteristics of these antigens, as well as their clinical implications and the value of studying them.

II Minor histocompatibility antigens (AgMHs)
II.1. General

Attention to minor histocompatibility (non-HLA histocompatibility) increased with the mention of its role in post-transplant complications of hematopoietic stem cells and solid organs, even in HLA-identical contexts. Historically, the discovery of minor histocompatibility antigens (AgMHs) or non-HLA antigens goes back to studies carried out by George Davis Snell in 1948, when he differentiated between minor and major histocompatibility (Snell.G,1948). Since then, the concept of minor histocompatibility has developed, but very slowly compared with the boom in the investigation of the HLA system and its associated genes. Minor histocompatibility antigens are alloantigens derived from polymorphic endogenous proteins and capable of generating alloimmune reactions between geno-identical HLA individuals (Goulmy.E, 1996; Spierings.E et al., 2007; Spencer.Ch et al., 2010). They are most often peptides of 8 to 11 amino acids, and their presentation is restricted to HLA class I and class II molecules (Goulmy.E, 1996) (Table I).

Table 1: Characteristics of hmaminor histocompatibility antigens (AgMHs) (Spencer.Ch et al.,2010)

AgMHs	Tissue distribution	Molecules HLA	Gene /chromosome	Peptides (variation)
HA-1	Hematopoieti c Solid tumors	HLA-A0201 HLA-B60	HMHA1/19p13	VLHDDLLEA Variation in position (3)and (6)
HA-2	Hematopoietic	HLA-A0201	MYO1F/7p12p13	VIGSVLISV Not described
HA-3	ubiquitous	HLA-A0101	LBC/15q24-25	VTEPGTAQY Variation in position (2)
HA-8	Ubiquitous	HLA-A0201	KIAA0020/9p.24	RTLDKVLEV Variation in position (1) and (9)
HB1	Hematopoietic	HLA-B4403	HMHB1/5q31-32	EEKRGSLHVW Variation at position (8)
ACC1	hematopoietic	HLA-A2402	BCL2A1/15q25.1	DLYQCVLOI Variation at position (8)
ACC2	Hematopoietic	HLA-B4403	BCL2A1/15q25.1	KEFEGIINW Variation on position (6)
UGT2B17	Hematopoietic	HLA-A0206 HLA-A2909 HLA-B4403	UGT2B17/4q13.2	CVATMI FMI AELLNIP FLY Deletion at position (2) and (10)
LRH1	Hematopoieti c Solid tumors	HLA-B0702	P2X5	TPNQRQNVC Polymorphism frame shift
HwA9	Hematopoietic	HLA-0301	SP110/2q37.1	SLPRGTSTPK Variation in position 4
PANE1	Hematopoietic (B Lymphoides)	HLA-0301	HwA10/22q13.31	RVWDLPGVLK SNP at the position (1)
HwA11	Solid tumors	HLA-A0201	C19orf48/19q13	CIPPDSLL FPA SNP
ECGF1	Solid tumors	HLA-B0702	ECGF1	RPHAIRRPLAL Variation in the position (3)
CTSH	Hematopoietic	HLA-A3101 HLA-A3303	CTSH/15q25.1	ATLPLLCAR WATLPLLCAR the second variant does not exist
ADIR	Hematopoietic ; solid tumors	HLA-A0201	ADIR/1q25.1	SVAPALALFPA SNP at position (9)
ACC6	Ubiquitous	HLA-B4403	HMSD/18q22	MEIFIEVFSHF variation in position (7)
HY-A1	Ubiquitous	HLA-A0101	DFFRY/Yq11.1	IVDCLTEMY Variation in position 4
HY-B7	Ubiquitous	HLA-A3303 HLA-B0702	SMCY /Yq11.1	FIDSYICQV Variation ? SPSVDKARAEL Change(8)
HY-A4	Ubiquitous	HLA-A3303	TMSB4Y/Yq11.2	EVLLRPGLHFR Variation?
HY-B60	Hematopoietic	HLA-B0801	UTY/Yq11.1	LPHNRTDL Variation in position 5
HY-B4	Hematopoietic solid tumors	HLA-B5201	RPS4Y/Yp11.3	TIRYPDPVI change in position(8)

AgMHs are peptides belonging to a variety of proteins whose cellular exposure is most often of the external membrane type. Genetically, these proteins are encoded by genes carrying allelic polymorphisms and located outside the classical HLA region, i.e. inherited independently of the HLA locus (Goulmy.E, 1996; Spencer.Ch et al., 2010).

The chromosomal location of these genes varies from autosomal to gonosomal (Perreault.C et al., 1998; Malarkannan.S et al., 2005) (Table I). In theory, any polymorphic protein of internal origin can be a candidate carrier for one or more AgMHs (Goulmy.E, 1996). Fundamentally, the basic genetic mechanism behind this group of antigens is undoubtedly genetic polymorphism, making it difficult to find perfect phenotypic identity between any two individuals, except in monozygotic or "identical" twins.

❖ **Nucleotide polymorphism**

Phenotypic diversity between individuals is generally attributed to variations in the nucleotide sequences that make up their genomes. Of the three billion bases making up the human genome, 99.9% are identical from one individual to another. The remaining 0.1% represent the variations or polymorphisms that underlie the differences between individuals. Around 90% of these variations are single nucleotide polymorphisms (SNPs), and 10% are either insertions/deletions or tandem repeats (VNTRs) (Ameziane.N et al ., 2005).At the gene level, these variations can be located either along the entire gene or in a single region, namely the coding, non-coding or regulatory region (Figure 4). A gene (locus) is said to be polymorphic if it exists in at least two forms (alleles) in any population, and the minor allele frequency (MAF) is greater than or equal to 1 (Ameziane.N et al.,2005).

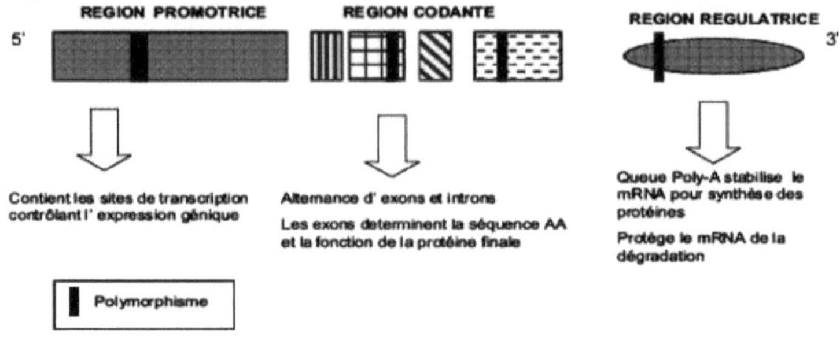

Figure 4: Possible locations of polymorphisms in a gene(Ameziane.etal.,2005)

II.2.1. Multi-nucleotide polymorphisms

Multi-nucleotide polymorphisms correspond to the existence of a variable number of repeats of a particular motif in a DNA sequence. These chain-repeated sequences are common in the genome. Depending on motif size and number of repeats, a distinction is made between satellites, minisatellites and microsatellites (Le Morvan.Vet al., 2005).

II.2.1.1 Single-nucleotide polymorphisms

II.2.1.1.1. RFLPs

RFLPs are point variations in DNA sequence revealed by changes in the restriction map. The size of DNA fragments varies after treatment with a restriction enzyme, hence the name RFLP (Restriction Fragment Lenght Polymorphism). They correspond to point mutations that abolish or create a restriction site. RFLPs were used to establish the first human genetic maps. These RFLP polymorphisms are gradually being replaced by SNPs, which are more numerous and easier to type and interpret.

II.2.1.1.2. SNPs

SNPs, pronounced "snips" or Single Nucleotide Polymorphisms, are the most frequently encountered genetic variants in the human genome. These polymorphisms are defined by the coexistence of at least two different bases at a given point in the genome, either by substitution (replacement of one nucleotide by another), deletion or insertion of a base (Collins.A et al.,1999). These SNPs are almost exclusively bi-allelic and constitute one of the major sources of inter-individual genetic and phenotypic variation. They are associated with inter-population diversity and differences in disease susceptibility. SNPs are distributed along the human genome at a density of around 1/300 base pairs (bp). These SNPs can be located at exons, introns, untranslated regions (UTRs) or intergenic regions (Doris.PA ,2002). Many of these SNPs, despite the variations they cause in the coding and/or regulatory parts of a gene, generally affect neither the function of the gene nor that of the encoded protein, and are therefore referred to as neutral polymorphisms. In some cases, however, these polymorphisms can lead to protein alteration through amino acid substitution (non-synonymous SNPs), or through altered splicing or transcriptional or post-transcriptional regulation, and are referred to as functional polymorphisms.

II.2.2 Nucleotide polymorphism and AgMHs

AgMHs are originally peptides belonging to a variety of externally-exposed endogenous transmembrane proteins. Genetically, these proteins are encoded by genes carrying mostly SNP-type allelic polymorphisms (Malarkannan et al., 2005). It has been suggested that a single protein of this type can contribute, after degradation, to the formation of one or more immunogenic peptides, and hence one or more AgMHs. It has also been shown that most of these antigens result from a single nucleotide polymorphism, but there are others that result from two polymorphisms, as in the case of the HA-1 antigen (Spencer.Ch et al., 2010). Generally, these antigens are the products of nonsense nucleotide substitutions (nonsynonymous SNPs) that cause amino acid changes in carrier proteins, as in the case of AgMHs HA-3, HA-8 etc. They are also the products of nonsense substitutions, such as PANE1, and of nucleotide deletions (Murata.M et al.,2003).

II.2.1. Mechanism of generation and presentation of AgMHs

Proteins are constantly renewed, depending on their importance and biological role, by biochemical processes under the supervision of the immune system. Indeed, the paradigm of self immunotolerance and non-self elimination relies heavily on the surveillance provided by cytotoxic T lymphocytes. This surveillance is made possible by a cellular protein processing system, which delivers peptide samples to HLA class I molecules expressed on the cell surface (Warren.D, 2007). To continuously produce a range of peptides representative of the world of proteins made by the cell, faithfully and in near-real time, this system exploits fundamental pathways of cellular metabolism, to which it associates certain actors with a more specialized function. For example, a key component of cellular protein metabolism, the proteasome (cytosolic protease), is exploited as a source of peptides, harvesting a small fraction of peptides for immune surveillance (Figure 5). At this level, endogenous proteins are degraded from the C-terminal region into peptides of 4 to 24 amino acids. Only a 15% fraction of these peptides will be of the right size, and can therefore be presented by HLA class I molecules to CD8 lymphocytes. The remaining peptides are either short or longer than the size required for antigenic presentation. The latter can be shortened on the NH2-terminal side by aminopeptidase activities present in the cytosol or endoplasmic reticulum. The peptides formed are subsequently translocated from the cytoplasm to the endoplasmic reticulum via 2 peptide transporters (Transporter properdine), TAP1 and TAP2, encoded by a gene located in the HLA region. Arriving at the endoplasmic reticulum, these peptides associate with HLA class I molecules organized in a large tetrameric complex. It is the association of the α-heavy chain, the β2-microglobulin and the peptide that enables the HLA molecule to reach the cell surface via the Golgi apparatus compartments. Here, the antigens bound to the ligands of the HLA molecules will be exposed at the cell surface to CD8 T lymphocytes. In the case of HSC transplantation, recognition of these peptides by donor T cells infused with the hematopoietic graft can, in the majority of cases, trigger immune responses that are usually fatal (Malarkannan.S et al., 2005 ; Spencer.Ch et al., 2010).

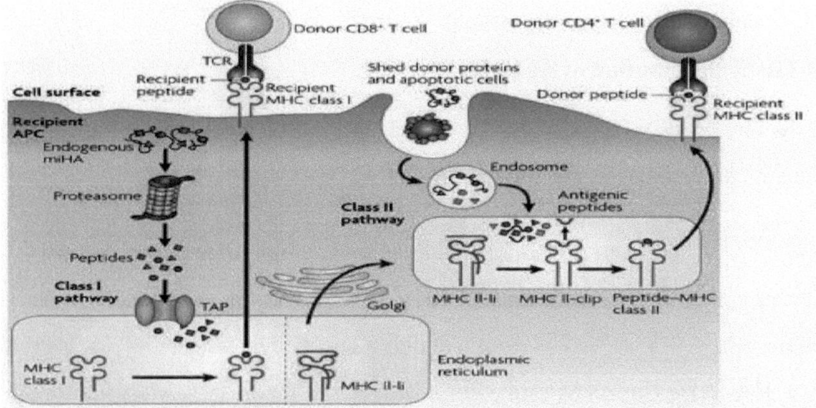

Figure 5: Generation and presentation of AgMHs by HLA class I/II molecules (Warren.D2007)

A second situation may arise in which the proteins are of exogenous origin. In this case, an internalization and degradation step is necessary to prepare the peptides for antigenic presentation. These peptides are then presented by HLA class II molecules, which are specialized in antigenic presentation to CD4 T lymphocytes. Overall, whatever the situation, AgMH-expressing APCs play a vital role in the process of immune alloreactivity.

II.2.2 Methods for identifying AgMHs

Identification of the tissue expressing a candidate minor antigen is a fairly complicated process requiring at least two steps (Malarkannan.S et al.,2005):

- A purification stage (biochemical study) in which peptides are extracted and purified from lymphocyte TCRs using several methods, the best known of which is high-performance liquid chromatography (HPLC) (Malarkannan.S et al., 2005).

- A deciphering step (genetic study) in which the amino acid sequences are sequenced and aligned with known sequences to determine candidate carrier proteins. This stage requires a blast with templates published in the PDB (Protein Database Bank) and GenBank (NCBI Gene Bank) databases.

- Once the candidate antigen has been identified, the next step is to localize the tissues expressing it.

II.3 Tissue distribution of AgMHs

For the few dozen minor antigens identified, specialized studies have shown that there are two types of tissue distribution (Table I):

➢ The first distribution is ubiquitous: minor antigens are present on every cell in the body. Examples i n c l u d e minor antigens encoded by the Y chromosome (HY), as well as certain antigens encoded by autosomes, such as HA-3, HA-8 (De Bueger.M et al., 1992; Brickner.A et al., 2001; Malarkannan.S et al., 2005).

➢ The second distribution is restricted to a tissue or group of cells, such as AgMHs from the HA-1, HA-2 HB-1 and PANE1 hematopoietic lineages. Distribution may also be restricted to tumor cells or cells treated with interferon gamma (INF-γ) or TNF factor (De Bueger.M et al., 1992; Dolstra.H et al., 1997; Warren.EH et al., 2002; Malarkannan.S et al., 2005).

Immunodominance of AgMHs is a concept recently introduced to express the difference between these antigens in terms of immunogenicity (Perreault et al., 1998; Shlomchik, 2007). It is a new paradigm in which the AgMHs group is subdivided into two subgroups:

➢ A first subgroup includes all antigens whose immunogenicity

is very strong, and their impact on the occurrence of certain complications is clearly evident. They are referred to as "minor immunodominant antigens", such as HA-1,HA-2,PECAM- 1,HA-3,HA-8,SP110 and PANE1 (Rufer et al.,1998;

Maruya et al.,1998 ; Spencer.Ch et al., 2010).

➢ A second sub-group comprises all minor antigens, whose immunogenicity is relatively low. The intensity of immune reactions involving these peptides is generally much lower than those involving immunodominant antigens.

II.4 Clinical implications of AgMHs
❖ **AgMHs and hematopoietic stem cell transplantation**

When HLA-identical HSCs are transplanted, two types of immune cells of different origins are brought together: host cells carrying the host's minor histocompatibility antigens, and graft cells carrying the donor's minor histocompatibility antigens (Figure 6). With the exception of true twin

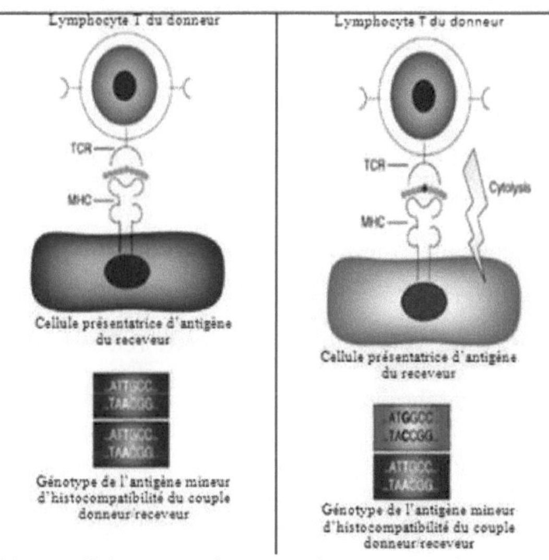

transplants, this cellular coexistence is most often accompanied by incompatibility at the level of AgMHs, which could be at the origin of two types of allogeneic reactions. The first type of reaction is host-versus-graft (HVG), classically known as "rejection". The second type is graft-versus-host disease (GVHD), in the case where the host is immunocompromised.

II.4.1. Host-versus-graft reaction or rejection (HVG)

Rejection or HVG (host versus graft effect) is a rare complication (less than 2%) that can occur following hematopoietic stem cell transplantation between geno-identical HLA individuals. This complication may be due to the persistence, after conditioning, of mature recipient T lymphocytes capable of activating and rejecting the graft (Falkenburg. J et al.,1991). These immunocompetent lymphocytes recognize minor histocompatibility antigens presented in association with HLA molecules on the surface of hematopoietic graft cells (Figure 7 and 8), triggering immune reactions to eliminate them (Pierce.R et al.,2001).

II.4.2. Graft-versus-host disease (GVHD)

It is clear that graft-versus-host disease (GVHD) is the most frequent phenomenon in HSC transplantation.Intensified by the condition of the recipient, who may be totally immunocompromised, and the presence of a mismatch in histocompatibility antigens between the transplanted pair, the GVH reaction can take two forms: a beneficial form or Graft-versus-Leukemia (GVL) or Graft-versus-Tumor (GVT), and a harmful form or graft-versus-host disease/reaction (Welniak.L et al.,2007).

a.Graft-versus-Tumor/Leukemia reaction (GVT/GVL)

The notion that alloreactivity may generate an anti-leukemic effect has been suggested in studies of leukemic mouse models where the rate of disease relapse after sygeneic transplantation is higher than that observed after allogeneic transplantation (Andrew.R ,2008).In humans, GVL or GVT is almost always observed in association with GVHD (Hambach.L and Goulmy.E ,2005).These notions have highlighted the possibility of using hematopoietic stem cell transplantation to treat certain malignant diseases. In this case, the use of the donor's mature T lymphocytes is necessary for the recognition of AgMHs presented by HLA molecules on the surface of the recipient's leukemia or tumor cells. This recognition could trigger immune reactions against these cells in order to eliminate them: this is the GVL or GVT effect (Figure 7 and 8) (Andrew. R ,2008).R ,2008).furthermore, Hambach's studies demonstrate that T lymphocytes specific to AgMHs could generate immune reactions directed against hematopoietic cells, resulting in the GVL effect without GVHD (Hambach.L et al., 2008; Hambach.L et al., 2007). This beneficial effect depends on the tissue distribution of the AgMHs: only AgMHs with a distribution restricted to hematopoietic or tumoral lineages are the major targets of this reaction (Falkenburg.J et al.,2004).

b.Graft-versus-host disease (GVHD)

This is the most serious complication of allogeneic hematopoietic stem cell transplantation, responsible for morbidity that is most often very severe, leading in 50% of cases to patient mortality (FerraraL.Jet al., 2008). This reaction is triggered by the graft's effector cells against the patient's own cells (FerraraL.Jet al., 2008). It corresponds to the recognition of allo antigen (major or minor) by the donor's T lymphocytes (Moalic.V and Ferec.C ,2006).

There are two clinical forms of GVHD: the first is the acute form, which occurs within 100 days of transplantation, and the second is the chronic form, which occurs beyond 100 days post-transplant (Moalic.V and Ferec.C ,2006).

The AgMHs group includes all immunogenic peptides capable of triggering predominantly cell-mediated immune reactions between geno-identical HLA individuals (Malarkannan.S et al.,2005; Ferrara.J et al.,2008; Paczesny.S et al.,2010). The importance of minor antigenic histocompatibility in the occurrence of GVHD during HLA geno-identical HSC transplantation has been widely demonstrated (Niederwisser.D et al.,1993; Goulmy.E et al.1996; Markiewicz.M et al.,2009; Ferrara.J et al., 2008; Paczesny.S et al.,2010; Choi.S and Reddy.P,2010). It has been suggested that the disparity between HSC recipient and donor for a limited number of "immunodominant" AgMHs (Figure 7 and 8), such as HA-1, HA-2 ,CD31, HY-A2,HY and HA-8, is sufficient for the development of GVHD effects and the destruction of cells carrying the aforementioned antigens (Behar.E et al.,1996; Goulmy.E et al.1996; Martin.P et al.,1997Tseng.L et al.,1999; Gallardo.D et al.,2001; Socie.G et al.,2001; Grumet.FC et al.,2001; Cavanagh.G et al.,2005; El-Chennawi.FA et al.,2006 ;Miklos.D et al.,2005 ; Akatsuka.Y et al.,2003On the other hand, other AgMHs are slightly associated with GVHD effects, such as HA-3 in the case of multiple mismatches between the grafted pair (Spierings.E et al.,2003 ; Spellman.S et al.,2009).

In addition, the tissue distribution of AgMHs and the presence or absence of the specific presenting HLA ligand have a major impact on the onset of GVHD. Generally speaking, AgMHs with ubiquitous distribution (HA-4; HA-5; PANE 1; CD31; HA- 8,H-Y etc.) are the candidate antigens capable of developing a GVHD effect following HSC transplantation (Goulmy.E et al.,1996 ;Milkos.D et al.,2005).

c. AgMHs and solid transplants

Rejection reactions to solid transplants (kidney, skin...) even in the context of HLA geno-identity have also been associated with disparity between donor and recipient of the target organ in certain autosome- or gonosome-encoded AgMHs (Heinold.A et al.,2008). Graft elimination is achieved by immunogenic peptides presented on the surface of the grafted organ, triggering a cascade of immune reactions directed by the recipient's T lymphocytes. (Figure 8) (Heinold.A et al., 2008).

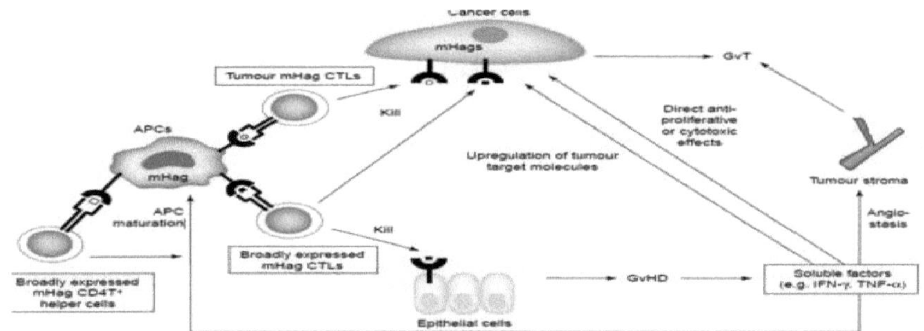

Figure 7: Clinical implication of AgMHs in post-HSC transplant reactions (Malarkannan.S et al.,2005)

Donor	Receiver	Clinical consequences
HLA disparity		Solid organ graft rejection
Identical HLA without AgMHs disparity		Grafting success
Identical HLA and disparity in AgMHs	HSCT or T cells	✓ Solid organ rejection ✓ HSC graft rejection ✓ GVHD ✓ GVL/GVT

Figure 8: Role of minor antigens in post-transplant reactions (Hambach.L and Goulmy.E ,2005)

❖ **AgMHs and fetomaternal tolerance**

D u r i n g pregnancy, alloimmunization can occur in both directions, i.e. from mother to fetus or from fetus to mother, in order to avoid abortions. This theoretical notion remains to be verified clinically (Goulmy.E ,2006).

❖ **Maternal/fetal alloimmunization**

During pregnancy, mother/fetal alloimmunization is directed against all minor fetal immunogenic antigens inherited from the father and lacking in the mother, encoded by the autosomes or by the Y chromosome (Figure 9) (Goulmy.E, 2006).

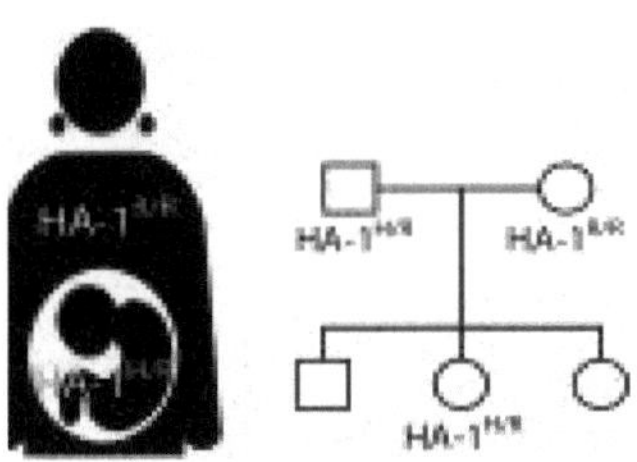

Figure 9: Disparity between mother and fetus for HA-1 AgMH as an example (Goulmy.E, 2006)

❖ **Fetal/maternal alloimmunization**

During pregnancy, fetus/mother alloimmunization can occur against AgMHs encoded only by autosomes. When the mother carries the immunogenic peptide of AgMHs not inherited by the fetus, the latter will develop immune reactions (Figure 10) (Goulmy.E, 2006).

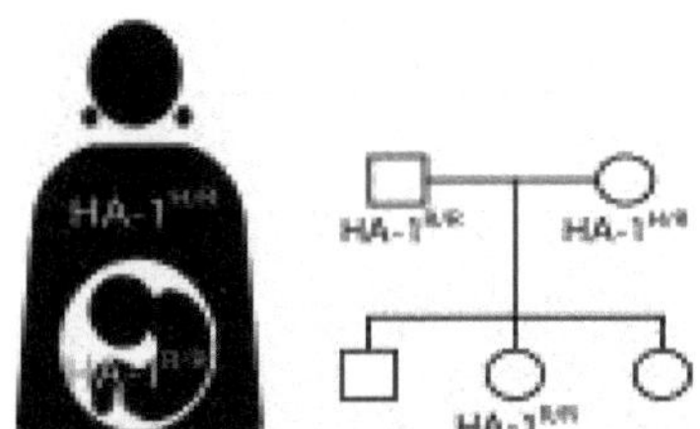

Figure 10: Disparity between mother and fetus for HA-1 AgMH as an example (Goulmy.E, 2006)

II.5. Inter-ethnic distribution of AgMHs

The allelic and genotypic distribution of AgMHs discovered varies from one ethnic population to another and from one AgMH to another (Tables II and III).

Table 2: Allelic distribution of AgMHs in different ethnic populations (Spierings et al.,2007)

gMHs/interethnic population		Asian/ Pacific (n=305)	Black (n=162)	White (n=2011)	Mexican (n=119)	Cap Color ed (n=65)	Mullato (n=123)
HA-1	H	47.6%	47.8%	35.9%	41.4%	35.2%	41.7%
	R	52.4%	52.2%	64.1%	58.6%	64.8%	58.3%
HA-2	V	91.9%	83.6%	75.6%	83.5%	84.9%	70.5%
	M	8.1%	16.4%	24.4%	16.5%	15.1%	29.5%
HA-3	T	57.4%	58.6%	65.2%	54.9%	55.6%	60.9%
	M	42.6%	41.4%	34.8%	45.1%	44.4%	39.1%
HA-8	R	39.7%	35.8%	45%	50.8%	29.4%	32.6%
	P	60.3%	64.2%	55%	49.2%	70.6%	67.4%
HB-1	H	69.9%	75.6%	74.2%	50.9%	73%	63%
	Y	30.1%	24.4%	25.8%	49 .1%	24%	37%
ACC-1	Y	42.5%	26.1%	26.8%	26.5%	36.7%	27.2%
	C	57.5%	73.9%	73.2%	73.5%	63.3%	72.8%
ACC-2	D	28.9%	6.5%	25.5%	11.6%	24.2%	15.2%
	G	71.1%	93.5%	74.5%	88.4%	75.8%	84.8%
SP110	R	76%	91.9%	61.2%	80.3%	74.2%	87%
	G	24%	8.1%	38.8%	19.7%	25.8%	13%
PANE1	R	71.2%	94 .7%	69.9%	82.5%	83.6%	63.3%
	R*	28.8%	8.4%	38.8%	19.7%	25.8%	13%
UGT2B17	+	44.9%	46%	81.3%	38.7%	24.3%	11.7%
	-	55.1%	54%	18.7%	61.3%	57.7%	88.3%

R*: Codon Stop

Table 3: Genotypic distribution of AgMHs in different ethnic populations (Spierings.E et al.,2007)

AgMHs/Interethnic population		Asian/ Pacific (n=305)	Black (n=162)	White (n=2011)	Mexican (n=119)	Cap Coloré (n=65)	Mullato (n=123)
HA-1	HH	24%	21.8%	13%	12.9%	7.8%	16.7%
	HR	47.2%	51.9%	45.8%	56.9%	54.7%	50%
	RR	28.8%	26.3%	41.2%	30.2%	37.5%	33.3%
HA-2	VV	84 .6%	69.8%	56.8%	69.6%	73%	54.5%
	VM	14.6%	27.8%	37.7%	27.7%	23.8%	31.8%
	MM	0.8%	2.5%	5.5%	2.7%	3.2%	13.6%
HA-3	TT	38%	38.9%	43.3%	31.3%	36.5%	26.1%
	TM	38.8%	39.5%	43.7%	47.3%	38.1%	69.6%
	MM	23.2%	21.6%	13%	21 .4%	25.4%	4.3%
HA-8	RR	16.1%	11.7%	19.7%	25%	11.1%	8.7%
	RP	47.1%	48.1%	50.4%	51.7%	36.5%	47.8%
	PP	36.8%	40.1%	29.8%	23.3%	52.4%	43.5%
HB-1	HH	50.2%	54.4%	53.7%	26.6%	54%	39.1%
	HY	39.4%	42.4%	41.2%	48.6%	38.1%	47.8%
	YY	10.4%	3.2%	5.2%	24.8%	7.9%	13%
ACC-1	YY	19 .1%	9.4%	7%	6.7%	9.4%	8.7%
	YC	46.9%	33.3%	39.6%	39.5%	54.7%	30.4%
	CC	34%	57.2%	53.5%	53.8%	35.9%	60.9%
ACC-2	DD	7%	0%	6.4%	0%	6.3%	0%
	DG	43.8%	13%	38.1%	23.2%	35.9%	30.4%
	GG	49.2%	87%	55.5%	76.8%	57.8%	69.6%
SP110	RR	60.3%	84.4%	37%	65.8%	53.1%	73.9%
	RG	31.5%	14.4%	48.3%	28.9%	42.2%	26.1%
	GG	8.2%	1.3%	14.7%	5.3%	4.7%	0%
PANE1	RR	49%	90.7%	47.2%	68.3%	70.3%	50%
	RR*	44.4%	8.1%	45.3%	28.3%	26.6%	27.3%
	R*R*	6.6%	1.2%	7.4%	3.3%	3.1%	22.7%
UGT2B17	+/+	20.1%	21.2%	66.1%	15%	5.9%	1.4%
	+/-	49.5%	49.7%	30.4%	47.4%	36.8%	20.6%
	-/-	30.4%	29.1%	3.5%	37.6%	57.3%	78%

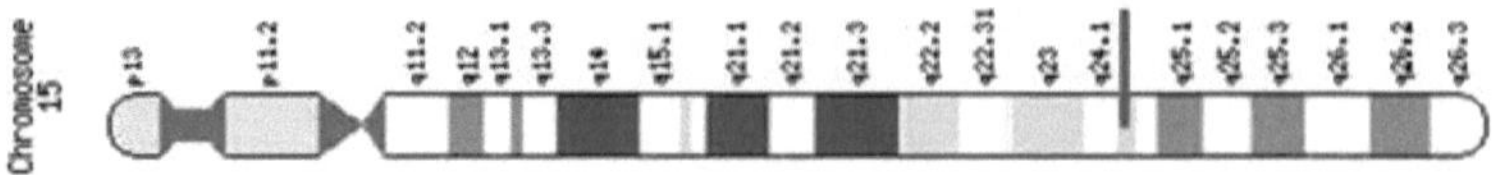

Figure 11: Location of the *AKAP lbc* gene on chromosome 15
<u>(ENSG00000170776)</u>

a.Structure of HA-3

AgMH HA-3 or (HA-3$^\text{T}$) is a peptide encoding the 2817 amino acid AKAP 13 (Protein Anchor kinase 13) protein. It's a 9-amino-acid peptide (**VTEPGTAQY**), stretching from amino acid 1216 to 1224 of the aforementioned protein. This peptide, highly immunogenic and containing a Threonine (Thr) at position 2, is generated at the level of proteasomes in a framework of classical protein metabolism. Another copy of this peptide, named HA-3$^\text{M}$ and containing a Methionine (Met) instead of Threonine 2, can also be produced at the cellular level, if the subject is positive for the *HA-3*T* allele. This second peptide is an inert peptide, since it is tolerated by T lymphocytes and known as part of the "Self" (Spierings.E et al., 2003).

b.HLA molecules from presentation

The immunogenic HA-3$^\text{T}$ peptide has greater affinity for HLA-A*0101 molecules present on the surface of CPAs. In contrast, HA-3$^\text{M}$ has less affinity for HLA-A*0101. This was demonstrated in Spierings' studies. Indeed, there is no good correlation between HA-3$^\text{M}$ and HLA- A*0101 due to the poor binding of methionine and HLA-A*0101 molecules. This polymorphism could be the origin of a GVHD reaction (Spierings.E et al.,2003 ; Shastri.N et al.,2002).

c.Distribution Tissue

The HA-3 minor antigen is ubiquitously distributed and its expression is unrestricted (Bradely .C et al., 2005).

OBJECTIVES

In this work, we have set ourselves the following objectives:

1. Perform a bioinformatics study to locate the polymorphic site on the *AKAP lbc* gene coding for the HA-3 minor antigen. In addition, check the primers chosen for molecular genotyping.

2. Develop a PCR-SSP technique for molecular genotyping of this site.

3. Determine allelic and genotypic frequencies and verify H.W equilibrium for this minor histocompatibility antigen.

4. Compare results with those reported in other populations.

MATERIAL & METHODS

MATERIALS AND METHODS

A. Bioinformatics study

I. Definition of bioinformatics

Bioinformatics is the discipline of analyzing biological information, mostly in the form of genetic sequences and protein structures.

In this study, we used this approach to verify the specific primers chosen to type the HA-3 target antigen, as cited in Spierings' article (E. Spierings et al., 2006).

II. Primer verification: Blasting step

II.1. Principle
This is a necessary step to ensure that the amplification process runs smoothly. This step is used to check the specific binding of primers to the nucleotide sequence of interest, in order to avoid non-specific or spurious amplifications.

Verification of these primers requires "blasting" against the known sequenced human genome, in order to see where they may have attached. Blasting can be carried out using the BLAST service provided by NCBI or the in Silico PCR service provided by UCSC Genome Bioinformatics.

II.2. Primer blasting

The primers chosen to genotype HA-3 (Table IV) were verified on :

http://genome.ucsc.edu: In the "UCSC in-Silico PCR" link (Figure 14 and 15), inserting the two primers in the "Forward Primer" and "Reverse Primer" boxes gave us :

- Primer localization on the human genome.

- Amplicon size
- The amplified sequence
- Primer temperatures (Tm)
- Primer size.

http://ncbi.nlm.nih.gov/blast: This site gives us all the possible ways in which the two primers could be attached to the human genome. These possibilities will be classified according to :

- ✓ Alignment scores.
- ✓ The identity of the primer in relation to the human genome, etc.

Table 4: Primers used for HA-3 alleles, (E.Spierings et al.,2006)

Primer pairs	Primer sequence	Primer temperature (Tm in °C)	Amplicon size
HA-3*C (Thr) **(Sense and anti-sense)**	5'CTTCAGAGAGACTTGGTCAC3' 3'GTTCATGAGCCCATGTTCCAT5'	50.8 62.1	129 bp
HA-3*T (Met) **(Sense and anti-sense)**	5'CTTCAGAGAGACTTGGTCAT3' 3'AGACTCAGCAGGTTTGTTAC5'	50.8 52	318pb

Figure 14: The Blasting site
(http://genome.ucsc.edu)

Home   Genomes   Blat   Tables   Gene Sorter   Session   FAQ   Help

UCSC In-Silico PCR

```
>chr15:86122635+86122763 129bp CTTCAGAGAGACTTGGTCA GTTCATGAGCCCATGTTCCAT
CTTCAGAGAGACTTGGTCAtggagccaggcacagcccagtattcctctgg
aggtgaactgggaggcatttcaacaacaaatgtcagtaccccagacactg
caggggaaATGGAACATGGGCTCATGAAC
```

Primer Melting Temperatures

```
Forward: 50.8 C cttcagagagacttggtca
Reverse: 62.1 C gttcatgagcccatgttccat
```
The temperature calculations are done assuming 50 mM salt and 50 nM annealing oligo concentration. The code to calculate the melting temp comes from Primer3.

Figure 15: Example of HA-3*C allele blasting (http://genome.ucsc.edu)

B. Molecular study

I.Study population

The study was carried out on 150 healthy, unrelated blood donors recruited from different regions of the country. All blood samples were taken on EDTA anticoagulant at the Centre National de Transfusion Sanguine de Tunis (CNTS), at a rate of 2 x 5 ml tubes per blood donation.

II. Molecular manipulations

II.1 Genomic DNA extraction

Throughout this work, DNA extraction was carried out using the Salting out method adapted to the conditions of the CNTS haematology laboratory (Miller et al., 1988).

II.1.1. Principle

Blood leukocytes are the major source of genomic DNA. The blood sample is subjected to a hypotonic solution causing lysis of the red blood cells. After centrifugation, the recovered white blood cell pellet is mixed with a white cell lysis solution designed to burst white blood cells, and a proteinase K solution to degrade membrane proteins. Protein release is carried out using high ionic strength (6 M NaCl). Finally, genomic DNA is extracted using an absolute ethanol solution.

II.1.2. How it works

II.1.2.1. Lysis of red blood cells

After deplasmatization, a hypotonic red cell lysis solution (RCLS) is added to the blood sample. After shaking, centrifugation at 2300 rpm for 15 min at room temperature and removal of the supernatant, the white blood cell pellet is lysed again. This red cell lysis step is repeated 3 to 4 times until a hemoglobin-free white cell pellet is obtained.

II.1.2.2. Lysis of white blood cells and protein precipitation

White pellet is supplemented with white lysis solution (WLS) and proteinase K. The volume of SLB added varies from 600µl to 1200µl, depending on the size of the pellet. Proteinase K, reconstituted to a concentration of 20 mg/µl, is added to the suspension at a rate of 20 to 35µl, depending on pellet size. After homogenization, the suspension is incubated at 42°C overnight or at 56°C for 3 hours. Next, 200µl of saturated NaCl (6M) is added and the suspension is vortexed until milky. After centrifugation for 30 min at 5000 rpm, the supernatant is collected in a new tube.

II.1.2.3. DNA precipitation

After addition of an equal volume of absolute ethanol (95°) cooled to -20°C and removal of polysaccharides, a DNA jellyfish is formed by gentle agitation of the suspension. This medusa is transferred to a clean eppendorf tube, washed at least 3 times with 70% ethanol to remove all traces of salt, and finally dried in a freeze-dryer (DNA plus).

II.1.2.4. DNA dissolution

After lyophilization, the dried DNA is solubilized in a volume of distilled water chosen according to the size of the jellyfish. This solubilization requires continuous stirring overnight at 42°C. After complete dissolution, the optical density (OD) of the resulting DNA solution is systematically measured at 260 nm and 280 nm to determine the concentration and purity of the extracted DNA.

II.1.2.5. DNA assay

It is rarely necessary to carry out a very precise assay, and in our case a simple estimate of concentration is sufficient. Relying on the fact that purine and pyrimidine bases absorb strongly in the ultra-violet at 260nm. It becomes possible to estimate the concentration of a soluble DNA solution. Knowing that one unit of optical density measured on a double-stranded DNA solution at 260nm corresponds to 50 ng/µl and that a 1/100 dilution is made[ème]. The approximate concentration of the DNA stock solution is calculated by the following formula :

[Parent DNA] = 100 x 50 (µg/ml) x A; with A: Do at 260nm.

II.1.2.6. Estimation of DNA purity

The purity of the DNA thus obtained is verified by performing a UV absorption spectrum at 260 nm and 280 nm on the 1/100th diluted DNA solution. Proteins absorb at 280nm, while nucleic acids absorb at 260nm. The degree of purity of a nucleic acid solution is estimated by calculating the ratio :

Ratio = OD 260 nm / OD 280 nm

The DNA solution is said to be of good quality when the ratio is between 1.8 and 2.

II.1.2.7. Dilution preparation

DNA stock solutions should be diluted in doubly distilled water and adjusted to 100 ng/µl, which is the optimum concentration for gene amplification.

II.2. AgMHs genotyping technique by PCR - SSP or ASA

Thanks to its performance, the PCR-SSP/PCR-ASA (Specific Allele Amplification or Specific Sequence Amplification) technique was chosen for molecular genotyping of the HA-3 minor histocompatibility antigen.

A key feature of this technique is its speed compared with other molecular techniques. In just a few hours, it can determine the allelic versions of HA-3 and its corresponding genotypes.

II.2.1 Principle of PCR-SSP

The PCR-SSP technique is based on the principle of amplification by Taq polymerase, and can only be effective when the primer is perfectly complementary to the target DNA sequence. Primer pairs are defined to be specific for a single allele or a group of alleles. Under very precise PCR conditions, the specific primer pair enables amplification of the target sequence (positive result), while non-complementary primer pairs give no amplification (negative result). After PCR, the amplified DNA fragments are separated by electrophoresis on an agarose gel and visualized under UV light after staining with ethidium bromide. Interpretation of PCR-SSP results is based on the presence or absence of a specific amplified fragment. Many factors can affect PCR efficiency (pipetting errors, poor DNA quality, presence of inhibitors, etc.), s o an internal control primer pair is incorporated into each PCR reaction. This control primer pair (sense primer : GCCTCCCCAACCATTCCCTTA and antisense primer: TCACGGATTTCTGTTGTG TTTC) amplifies a conserved region of the human HGH gene that is present in all DNA samples, enabling the integrity of the PCR reaction to be verified. In the case of allele-specific amplification of an antigen (positive band), the internal control band may be weak or absent due to competition between primer pairs for PCR ingredients (Taq; dNTPs etc.).

II.2.1   Development of PCR-SSP applied to AgMHs *genotyping*

Several amplification tests were carried out during the course of this work. The first tests were carried out under standard PCR conditions and at a temperature of 61°C. Aspecific amplification bands were obtained. Successive increases in hybridization temperature to 62.5° and adjustment of the HGH concentration solved the problem of aspecific amplifications. The final PCR conditions are summarized in Tables V, VI and VII.

II.2.2   Thermocycler program

For allele-specific amplification of the antigen under study, we optimized the concentrations of the various ingredients. At the end of the tests, a single program was used for the alleles, which will facilitate genotyping in the laboratory. The various amplification steps are summarized in Table VIII.

II.2.3   Amplification control

Amplified DNA fragments are separated by electrophoresis for 20 min on a 2% agarose gel and visualized by staining with ethidium bromide (BET). Interpretation of PCR-SSP results is based on the presence or absence of specific amplified fragments.

Table 5: Final PCR conditions for HA-3*C/HA-3*T allele genotyping

Ingredients	CC° initial	CC°/PCR	V/PCR
EBD	-	-	1.7 μl
Buffer (Promega)	5x	1x	2μl
MgCl 2(Promega)	25mM	1.5mM	0.6 μl
dNTPs	2mM	0.2mM	1 μl
Primer HGH sens (700bp)	2000nM	160nM	0.8μl
HGH anti-sense primer (700bp)	2000nM	160nM	0.8 μl
Specific sensory primer	2000nM	200nM	1 μl
Specific anti-sensitivity primer	2000nM	200nM	1 μl
Taq polymeras e(Promeg)	5U/μl	0.5U	0.1 μl
Genomic DNA	100ng/ul	100ng	1 μl
Final volume	-	-	10 μl

Table 8: PCR program used for HA-3 amplification

Step		Temperature	Duration	Number of cycles
Denaturat ion		94° C	5 min	1
Amplification	Denaturatio n	94° C	30 secs	30
	Hybridizati on	62.5°C	40 secs	
	Elongation	72° C	30 secs	
Final elongation		72C	10 min.	1

C. Biostatistical study

I.Determination of allele frequencies

The allelic frequencies were calculated in two ways:

a.   *Manual calculation*

Allelic frequency = [2(n1 or n2) + n3] / 2N

With :
n1: number of homozygous A/A subjects
n2: number of homozygous B/B subjects
n3: number of AB heterozygotes

N: total population

b. Computerized calculation: This calculation was performed using Thesias software version 3.1.

II.Determination of genotype frequencies

Genotypic frequencies were calculated by simple counting.

Frequency. Genotypic = n /N; where n: number of subjects with the same genotype N: total number of subjects

III.   *Confidence interval*

A confidence interval, at the risk of error α, is an interval that has a probability 1-α of containing the true (unknown) value of a given parameter.

For a percentage, the confidence interval (CI) (at 95%, with a 5% risk of error) is calculated as follows:

$$[pi, ps] = p \pm 1.96 \sqrt{pq} / N$$

Where N: sample size

p: set frequency

q : 1-p

pi: lower limit of CI

ps: upper bound of CI

This interval allows us to say that for a sample of N, the results are reliable to within ± CI for a net result at a percentage error of 5%.

IV.   Hardy Weinberg equilibrium

This law is used to study population equilibrium over generations for the HA-3, HA-8 and PANE1 loci. Checking equilibrium according to this law is done as follows:

$$(p+q)^2 = p^2 + 2pq + q^2 = 1$$

Where p: frequency of allele x; q: frequency of allele y; $p + q = 1$

V.   Chi2 test of H.W

The $\chi 2$ goodness-of-fit test is used to estimate the probability of simultaneous observation of a series of deviations between observed and expected values.

$$\chi 2 \text{ test: } \Sigma \text{ (Obs -Tea) 2/ Tea}$$

Where **Obs: observed value; The: theoretical value**

VI. Degree of freedom

The number of variables in the Hardy-Weinberg test is not simply the number of phenotypes - 1 as in the χ^2 test of classical Mendelian proportions. The number of observed variables (number of phenotypes: k) is smaller in the test for conformity of allele frequencies to a Hardy-Weinberg distribution. This is because additional variables such as the number of alleles (r) are added. Thus, the combined number of degrees of freedom will be :

$$\textbf{ddl} = \textbf{(k-1)-(r-1)} = \textbf{k-r} = \text{number of phenotypes - number of alleles.}$$

RESULTS

RESULT S

The results of the molecular analysis reveal the presence of all the allelic versions of the gene studied.

Molecular analysis of the *AKAP lbc* gene involved the entire target population. It should be remembered that the target polymorphism is characterized by a single nucleotide substitution (SNP). An example of this genotyping is shown in Figure 16. The electrophoretic profile clearly shows the presence of 3 types of bands:

- ✓ A band size of 129 base pairs (bp) corresponds to the amplification of the immunogenic HA-3 *C allele (HA-3).^{T}
- ✓ A band of 318 base pairs corresponds to the amplification of the non-immunogenic HA-3 *T allele (HA-3).^{M}
- ✓ A 709 bp band corresponds to the amplification of the growth hormone (HGH) gene, which is the internal control.

Two profiles were obtained for this antigen in this study:

- ✓ A subject homozygous for HA-3*C/ *HA-3*C*

- ✓ A heterozygous subject HA-3*T/ *HA-3*C*

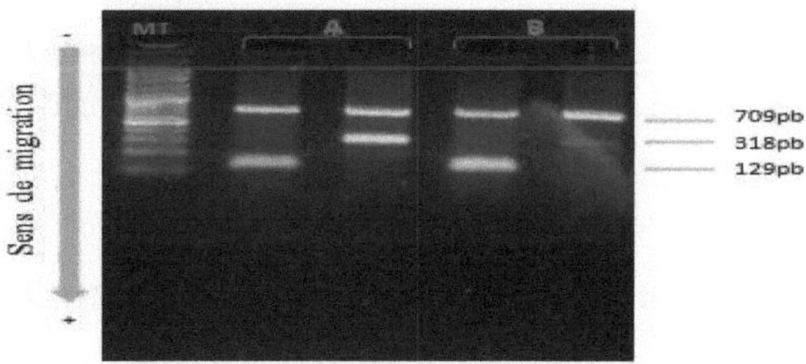

MT: marker size 100bp; A: subject heterozygous HA-3*C/HA-3*T; B: subject homozygous HA-3*C/HA-3*C

Figure 16: Eletrophoretic profile of the AKAP lbc gene on a 2% agarose gel

The statistical study showed that the *HA-3*C* allele is more frequent than the *HA-3*T* allele, with frequencies of 90% (±*0*.0480099) and 10% (± 0.0480099)

respectively (Table IX).

Genotypically (Table X), the majority of the population studied was homozygous for *HA-3*C / HA-3*C*. This genotype is predominant, with a frequency of 80%, followed by the heterozygous *HA-3*T / HA-3*C* genotype, which occurs in only 20% of the population studied. The same study showed the complete absence of the homozygous *HA-3*T / HA 3*T* genotype.

The study of *AKAP lbc* locus equilibrium was carried out after checking that our population met all the application conditions. The χ test^2 is applied for both allelic versions HA-3 **C* and *HA-3*T*. Table XI shows the observed and theoretical values calculated from the observed frequencies for this locus.

Hardy-Weinberg equilibrium analysis for the *AKAP lbc* locus coding for HA-3 shows that the Tunisian population is in equilibrium since the χ^2 cal value is lower than the theoretical χ^2 , and the difference is non-significant (p=0.173)>0.05

Table 9: Allelic frequencies of the AKAP lbc gene in the Tunisian population

Alleles	Frequency (%)	Confidence interval (IC)
HA-3*C*(Thr)*	90	± 0.04800099
HA-3 *T*(Met)*	10	± 0.04800099

Table 10: Genotypic frequencies of AKAP lbc in the Tunisian population

Genotype	Number	Frequency (%)
*HA-3*C / HA-3*C*	*120*	*80*
HA-3*C/ *HA-3*T*	*30*	*20*

Table 11: Observed and theoretical values of the AKAP lbc gene and calculation of $\chi2$

Genotypes	Observed values		Theoretical values		χ^2
	Number	**Frequencies**	**Number**	**Frequencies**	
HA-3*T/HA-3 *T	0	0	1.5	0.01	χ^2 calculated=1.848
*HA-3*C* /**HA-3 *T**	30	0.2	27	0.18	
*HA-3*C* /**HA-3 *C**	120	0.8	121.5	0.81	χ^2 theoretical=3.84

DISCUSSION

DISCUSSION

Minor histocompatibility antigens are immunogenic polymorphic endogenous peptides initially identified following post-HSC transplant complications between geno-identical HLA individuals. These AgMHs are presented on the cell surface in association with HLA molecules. Post-transplant complications such as GVHD and rejection are the result of a disparity between donor and recipient at the level of these AgMHs, triggering a series of immune reactions. Nevertheless, this disparity can induce beneficial effects such as the graft's reaction against tumor cells (GVT) or leukemia cells (GVL).

The study of the polymorphism and molecular basis of minor histocompatibility antigens in the Tunisian population is one of the research priorities of the Immuno-hematology laboratory of the Centre National de Transfusion Sanguine, and our work falls within this framework. We have therefore chosen to study the molecular polymorphism of the HA-3 antigen in our population. HA-3 is encoded by the autosomal *AKAP lbc* gene located o n chromosome 15. In the present work, we have developed a very rapid genotyping method to study the molecular polymorphism of the three above-mentioned genes and determine their allelic and genotypic frequencies in the Tunisian population.

Our study of a control cohort of healthy, unrelated individuals showed that :

➢ For the HA-3 antigen: the majority of the Tunisian population carries the immunogenic peptide HA-3$^\text{T}$ in both homozygous and heterozygous forms.

The homozygous *HA-3*C* / *HA-3*C* genotype is more frequent, with a frequency equal to 80%. The Hardy-Weinberg equilibrium study showed that the Tunisian population is in equilibrium for the *AKAP lbc* gene coding for this antigen. Compared with six ethnic groups cited in Spierings' article (Spierings E, 2007), the frequencies found for the HA-3 minor histocompatibility antigen are close to those reported in Europeans and African-Americans, but this antigen is not very polymorphic in our population, since the HA-3^T peptide is very frequent (Table XVIII and XIX).

Table 18: Comparison of HA-3 allele frequencies of six ethnic groups with those of the Tunisian population (Spierings E et al., 2007)

Population /allele	Tunisinne (n=150)	Asian (n=305)	Black (n=162)	White (n=2011)	Mexican (n=119)	Colored cape (n=65)	Mulatto (n=23)
*HA-3*C* (HA-3^T)	90%	57.4%	58.6%	65.2%	54.9%	55.6%	60.9%
*HA-3*T (HA-8)^M*	10%	42.6%	41.6%	34.8%	45.1%	44.4%	39.1%

Table 19: Comparison of HA-3, HA-8 and PANE1 genotype frequencies of six ethnic groups with those of the Tunisian population (Spierings E et al., 2007).

Population/ Genotype	Tunisian (n=150)	Asian (n=305)	Black (n=162)	White (n=2011)	Mexican (n=119)	Colored cape(n=65)	Mulatto (n=123)
*HA-3*TT*(HA-3)^{MM}	_%	23.2%	21.6%	13%	21.4%	25.4%	4.3%
HA-3*CT*(HA-3)^{MT}	20%	38.8%	39.5%	43.7%	47.3%	38.1%	69.9%
HA-3*CC*(HA-3)^{TT}	80%	38%	38.9%	43.3%	31.3%	36.5%	26.6%

This fundamental study was also necessary to predict the rate of disparity between geno-identical HLA individuals at a given minor antigen. Although the locus is in Hardy-Weinberg equilibrium, a n d based on the aforementioned allelic and genotypic data, we can deduce that the HA-3 antigen (individuals homo- and heterozygous for the immunogenic *HA-3*C* allele (HA-3^T)) is frequent in our population.

Clinically, the importance attached to the HA-3 antigen can be explained by the frequency of allogeneic reactions that can be triggered following recognition of each antigen by alloreactive T lymphocytes, in subjects positive for the antigen-specific HLA ligand. In this respect, it should be remembered that HLA-A*0101 is the specific molecule required for the presentation of HA-3 AgMHs to immunocompetent alloreactive lymphocytes.In the case of HSC allografts or HLA geno-identical solid organ transplants, the HA-3 immunogenic peptide^T presented by the HLA-A*0101 molecule is expressed in all Tunisians, which reduces the risk of developing immune reactions against this antigen in the case of hematopoietic stem cell allografts.

This study led us to investigate the probability (P) of a hematopoietic stem cell recipient carrying the immunogenic HA-3 allele.
specific HLA presenter molecule (Ayed et al., 2004) of the antigen, to the donor's cytotoxic T lymphocytes (Table XX, XXI and XXII).

The results show that out of 100 HSC recipients, 10 carry the immunogenic peptide HA-3^T and the HLA* molecule A0101 (Table XX).

Table 20: Calculation of the probability of recipients carrying HLA-A*0101/HA-3

<table>
<tr><td colspan="2" align="center">The TCSH context</td><td colspan="2" align="center">Receiver's HA-3 statement</td></tr>
<tr><td colspan="2"></td><td align="center">HA-3^T (+) (A)</td><td align="center">HA-3^M (-) (B)</td></tr>
<tr><td rowspan="2">HLA-A*0101 status
of the recipient</td><td>HLA-A*0101(+)
(C)</td><td align="center">0.1008

(=p(A)*p(C))</td><td align="center">0.0112

(=p(B)*p*(C))</td></tr>
<tr><td>HLA-A*0101(-)
(D)</td><td align="center">0.7999

(=p(A)*p(D))</td><td align="center">0.0888

(=p(B)*p*(D))</td></tr>
</table>

N.B. The frequency of the HLA-A*0101(+) allele in Tunisians is 0.112 (Ayed et al., 2004).

CONCLUSION AND OUTLOOK

CONCLUSION

The present study provided us with a PCR-SSP technique, and an estimate of the allelic and genotypic distribution of the *AKAP lbc* gene encoding the HA-3 antigen in a sample of the Tunisian population.

For this antigen, the most frequent allele is *HA-3*C* with a value equal to 90%. The genotype most present in our population is the homozygous *HA-3*C/* HA-3*C genotype (80%), followed by the heterozygous *HA-3*T/* HA-3*C genotype (20%). These fundamental findings will serve as a data base for future studies of the association between the occurrence of GVHD and the HA-3 minor histocompatibility antigen in Tunisian hematopoietic stem cell transplant recipients.

OUTLOOK

In the present study, we were able to create a Tunisian database for AgMHs.

This work deserves to be completed at a later date by a study of the association between the disparity for AgMHs for geno-identical HLA transplanted couples and carriers of specific HLA-A, for each antigen and the development of post-transplant hematopoietic stem cell reactions such as GVHD, rejection and GVL.

This study, followed by other fundamental and clinical studies on other AgMHs, could help transplant physicians to improve the results of hematopoietic stem cell transplantation in Tunisia, in order to offer better safety for transplanted patients and thus reduce post-transplant complications, especially the GVHD effect.

REFERENCES BIBLIOGRAPHICAL

References

Akatsuka Y, Warren EH, Gooley TA, et al.(2003) Disparity for a newly identified minor histocompatibility antigen, HA-8, correlates with acute graft-versus-host disease after haematopoietic stem cell transplantation from an HLA-identical sibling. *Br J Haematol* ;123: 671-5.

Ameziane N, Bogard M, Lamori Jl.(2005) Principes de biologie moléculaire en biologie clinique. Paris, *Elsevier Masson: p45-50.*

Andrew R,Rezvani,Rainer F.(2008) Separation of graft-vs-tumor effects from graft-vs -host disease in allogenic hematopoietic cell transplantation. *Autoimmunity; 30:172-179.*

Ayed K ,Ayed-Jendoubi S,Sfar I et al.(2004) HLA class I and HLA class II phenotypic ,gene and haplotypic frequencies in Tunisians by using molecular typing data. *Tissue Antigens*; 64:520-32.

Behar E, Chao NJ, Hiraki DD, et al.(1996) Polymorphism of adhesion molecule CD31 and its role in acute graft-versus-host disease. *N Engl J Med*; 334: 286-91.

Bierie B, Edwin M, Melenhorst JJ, Hennighausen L.(2004) The proliferation associated nuclear element (PANE1) is conserved between mammals and fish and preferentially expressed in activated lymphoid cells .*Gene Expr Patterns* ;4:389-95.

Bradely C. Pietz , Melissa B. Warden, Brian K. Duchateau ,and Thomas M. Ellis.(2005) Multiplex Genotyping of human minor histocompatibility antigens. *Human Immunology;* 66:1174-1182.

Brickner Anthony .G, Anne M. Evans, Jeffrey K. Mito, Suzanne M. Xuereb, Xin Feng, TetsuyaVictor H. Engelhard, Stanley R. Riddell and Edus H. Warren Nishida, Liane Fairfull, Robert E. Ferrell, Kenneth A. Foon, Donald F. Hunt, Jeffrey Shabanowitz. Ferrell, Kenneth A. Foon, Donald F. Hunt, Jeffrey Shabanowitz.(2006) The PANE1 gene encodes a novel human minor histocompatibility antigen that is selectively expressed in B-lymphoid cells and B-CLL. *Blood*; 107: 3779-3786.

Brickner Anthony .G , Edus H. Warren, Jennifer A. Caldwell, Yoshiki Akatsuka,Tatiana N. Golovina, Angela L. Zarling, Jeffrey Shabanowitz, Laurence C. Eisenlohr, Donald F. Hunt, Victor H. Engelhard, and Stanley R. Riddell .(2001) The Immunogenicity of a New Human Minor Histocompatibility Antigen Results from Differential Antigen Processing .*J Exp Med* ; 193(2): 195-206.

Carosella ED, Paul P, Moreau P, et al.(2000) HLA-G and HLA-E: fundamental and pathophysiological aspects. *Immunol Today*; 21: 532-4.

Cavanagh G, Chapman C, Carter V, et al.(2005) Donor CD31 Genotype Impacts on Transplant Complications after Human Leukocyte Antigen-Matched Sibling Allogeneic Bone Marrow Transplantation .*Transplantation* ; 79: 602-5.

Cesbron-Gautier A, Gagne K, Retière C, et al (2007) HLA system. EMC, *Hematologie*; 13-000-M-53, Doi: 10.1016/S1155-1984(07)47158-8.

Choi S, Reddy P. (2010) Graft-versus-host disease. *Panminerva Med* ;52: 111-24

Collins A, Lonjou C, Morton NE (1999) Genetic epidemiology of single-nucleotide polymorphisms.*Proc Natl Acad Sci USA*; 96: 15173-7.

Colombani J. (1993) In: HLA: Fonctions immunitaires et applications médicales. Paris, *John Libbey Eurotext*: p23-30.

Dausset J. (1958) Iso-leuco-antibodies. *Acta Haemat;* 20:156-66.

De Bueger M, Bakker A, Van Rood JJ, et al. (1992) Tissue distribution of human minor histocompatibility antigens.Ubiquitous versus restricted tissue distribution indicates heterogeneity among human cytotoxic T lymphocyte-defined non-MHC antigens. *J Immunol*; 149: 1788-94.

Dolstra H, Fredrix H, Preijers F, et al.(1997) Recognition of a B cell leukemiaassociated minor histocompatibility antigen by C.T.L. *J Immunol* ; 158:560-5.

Doris PA. (2002) Hypertension genetics, single nucleotide polymorphisms, and the common disease: common variant hypothesis. *Hypertension*; 39:323-31.

El-Chennawi FA, Kamel HA, Mosaad YM, et al.(2006) Impact of CD31 mismatches on the outcome of hematopoeitic stem cell transplant of HLA identical sibling. *Haematology*; 11: 227-34.

Falkenburg F MD, Willemze R MD.(2004) Minor Histocompatibility Antigens as targets of cellular immunotherapy in leukemia.*Best Practice and research clinical haematology;*17:415-425.

Falkenburg F ,Gosehnk H,van der harts D,Van Luxemburg-Heijs SAP ,Kooij-winkelaar YMC ,FABER LM,DE Kroon J,Brand A,Fibbe WE,Willemze R, and GoulmyE.(1991) Growth Inhibition of donorgemc leukemic precursor cells by histocompatibility antigen-specific cytotoxic T lymphocytes .*J Exp Med*; 174:27.

Ferrara JLM, Levine JE, Reddy P, et al (2008) Graft-versus-host disease. *The lancet;* 373: 1550-61.

Gallardo D, Arostegui JI, Balas A et al (2001) Disparity for the minor histocompatibility antigen HA-1 is associated with an increased risk of acute graft-versus-host disease (GVHD) but it does not affect chronic GVHD incidence, disease-free survival or overall survival after allogenic human leukocyte antigen-identical sibling donor transplantation. *British Journal of Haematology*; 114: 931-36.

Goulmy E. (2006) Minor histocompatibility antigens: from transplantation problems to therapy of cancer. *Hum Immunol*: 67: 433-8.

Goulmy E. (1996) Human minor histocompatibility antigens. *Curr Opin Immunol*; 8: 75-81.

Goulmy E, Schipper R, Pool J, et al.(1996) Mismatches of Minor Histocompatibility Antigens Between HLA-Identical Donors and Recipients and the Development of Graft-Versus-Host Disease After Bone Marrow Transplantation. *The New England Journal of Medicine*; 334: 281-5

Grumet FC, Hiraki DD, Brown BW, et al (2001) CD31 Mismatching Affects Marrow Transplantation Outcome. *Biology of Blood and Marrow Transplantation*; 7: 503-12.

Hajjej A,Hmida S,Kaabi H,Dridi A,Jridi A,El Gaaled A,Boukef K.(2006) HLA genes in Southern Tunisians (Ghannouch area) and their Relationship with other Meditterraneans

European Journal of Medical Genetics;49:43-56.

Hajjej A, Sellami MH, Kaabi H, et al.(2010) HLA class I and class II polymorphisms in Tunisian Berbers. *Ann Hum Biol; 38*(2): 156-164.

Hambach Lothar & Goulmy Els .(2005) Immunotherapy of cancer trough targeting of minor histocompatibility antigens. *Current opinion in Immunology*; 17:202-210.

Hambach Lothar , Spierings Eric and Els Goulmy. (2007) Risk assessment in haematopoetic stem cell transplantation: Minor Histocompatibility Antigens. *Best practice et research clinical hematology; 20*(2):171-187.

Hambach Lothar, Vermeij Marcel , Andreas Buser ,Zohara Aghai, Theodorus van der kwast and Els Goulmy.(2008) Targeting a single mismatching minor histocompatibility antigen with tumor-restricted expression eradicates human solid Tumors . *Blood*; 112:1844-1852.

Heinold A, OS,Ruhenstroth A,Laux G,Doehler B and Tran TH.(2008) Role of Minor Histocompatibility Antigens in Renal Transplantation. *American Journal of transplantation; 8:95-102.*

Klein J, Sato A.(2000) The HLA system. First of two parts. *N Engl J Med*; 343: 702-9.

Le Morvan V, Formento JL, Milano G, et al. (2005) Techniques for searching for genetic polymorphisms. *Oncology*; 7: 7-16.

Malarkannan Subramaniam ,Jeyarani Regunathan and Angela M. Timler.(2005) MinorHistocompatibility Antigens: Molecular targets for immunomodulation in tissue transplantation and tumor therapy. *Clinical and Applied Immunology*; 5(2):95-109.

Markiewicz M, Siekiera U, Karolczyk A, et al. (2009) Immunogenic disparities of 11 minor histocompatibility antigens (mHAs) in HLA-matched unrelated allogeneic hematopoietic SCT. *Bone Marrow Transplantation; 43:293-300.*

Martin PJ.(1997) How much benefit can be expected from matching for minor antigens in allogeneic marrow transplantation?.*Bone Marrow Transplantation*; 20: 97-100.

Maruya E, Saji H, Seki S, et al (1998) Evidence that CD31, CD49b, and CD62L are immunodominant minor histocompatibility antigens in HLA identical sibling bone marrow transplants. *Blood*; 92: 2169-76.

Miklos DB, Kim HT, Miller KH, Guo L, Zorn E, Lee SJ et al. (2005) Antibody responses to H-Y minor histocompatibility antigens correlate with chronic graft-versus-host disease and disease remission. *Blood*; 105: 2973-8.

Miller S, Dykes D, and Polesky H. (1988)A simple salting out procedure for extracting DNA from human nucleated cell. *Nucleic Acids Research* ; 16:1215-1218.

Moalic V, Ferec C. (2006) Graft-versus-host disease. *Pathologie Biologie*; 54: 304-308.

Murata Makoto, Edus H. Warren, and Stanley R. Riddell.(2003) A Human Minor Histocompatibility Antigen Resulting from Differential Expression due to a Gene Deletion. *J Exp Med*; 197(10):1279-1289.

Niederwieser D, Grassegger A, Auböck J, et al.(1993) Correlation of minor histocompatibility antigen-specific cytotoxic T lymphocytes with graft-versus-host disease status and analyses of tissue distribution of their target antigens. *Blood*; 81: 2200-8.

Paczesny S, Hanauer D, Sun Y, et al.(2010) New perspectives on the biology of acute GVHD. *Bone Marrow Transplant*; 45: 1-11.

Perreault C, Roy DC, Fortin C.(1998) Immunodominant minor histocompatibility antigens: the major ones. *Immunol Today*; 19: 69-74.

Pierce RA, Field ED, Mutis T, et al.(2001) The HA-2 minor histocompatibility antigen is derived from a diallelic gene encoding a novel human class I myosin protein .*J Immunol*; 167: 3223-30.

Rufer N, Wolpert E, Helg C, et al. (1998) HA-1 and the SMCY-derived peptide FIDSYICQV (H-Y) are immunodominant minor histocompatibility antigens after bone marrow transplantation. *Transplantation*; 66: 910-6

Schwartz BD.(1994) HLA class II transgenic mice: the chance to unravel the basis of HLA class II associations with disease. *J Exp Med*;180: 11-13.

Sellami MH, Kâabi H, Midouni B, et al.(2008) Duffy blood group system genotyping in an urban Tunisian population. *Annals of Human Biology*; 35: 406-15

Shastri N,Schwab S,Serwold T.(2002) Producing nature's gene-chips: the generation of peptides for display MHC class I molecules. *Annu Rev Immunol; 20:463-493.*

Shlomchik WD.(2007) Graft-versus-host disease. *Nat Rev Immunol*; 7: 340-52.

Snell GD.(1948) Methods for the study of histocompatibility genes. *Genet*; 19: 87-108.

Socie G, Loiseau P, Tamouza R et al.(2001) Both genetic and clinical factors predict the development of graft-versus-host disease after allogenic haematopoietic stem cell transplantation. *Transplantation*; 72: 699-706.

Spellman S, Warden MB, Haagenson M, et al.(2009) Effects of Mismatching for Minor Histocompatibility Antigens on Clinical Outcomes in HLA-Matched, Unrelated Hematopoietic Stem Cell Transplants. *Biology of Blood and Marrow Transplantation*; 15: 856-63.

Spencer CH, Gilchuk P, Dragovic SM, et al (2010) Minor histocompatibility antigens: presentation principles, recognition logic and the potential for a healing hand .*Curr Opin Organ Transplant*; 15: 512-25.

Spierings E, Brickner EG, Caldwell JA, et al.(2003) The minor histocompatibility antigen HA-3 arises from differential proteasome-mediated cleavage of the lymphoid blast crisis (Lbc) oncoprotein. *Blood*; 102: 621-9.

Spierings E, Hendriks M, Absi L, et al.(2007) Phenotype frequencies of autosomal minor histocompatibility antigens display significant differences among populations.*PLoS Genet* ;3: 1108-19.

Spierings Eric ,Jos Drabbels, Mathhijs Hendriks,Jos Pool,Marijke Spruyt-Gerriste,Frans Claas ,and Els Goulmy .(2006) A Uniform Genomic Minor Histocompatibility Antigen Typing Methodology and Database Designed to Facilitate Clinical Applications. *Plos One;* 1:1-8.

Sterpetti Paola, Andrew A. Hack, Mariam P. Bashar, Brian Park, Sou-De Cheng, Joan **H. M. Knoll, Takeshi Urano, Larry A. Feig and Deniz Toksoz.(1999)** Activation of the Lbc Rho Exchange Factor Proto-Oncogene by Truncation of an Extended C Terminus That Regulates Transformation and Targeting. *Mol. Cell. Biol*; 19(2): 1334-1345.

Terasaki PI, McClelland JD et al (1964) Microdroplet assay of human serum cytotoxins. *Nature*; 204: 998-1000.

Thomas ED, Lochte HL Jr, Lu WC, et al.(1957) Intravenous infusion of bone marrow in patients receiving radiation and chemotherapy .*N Engl J Med*; 257: 491-6.

Tseng LH, Lin MT, Hansen J.A et al.(1999) Correlation between disparity for the minor histocompatibility antigen HA-1 and the development of acute graft-versus-host disease after allogenic marrow transplantation. *Blood*; 94: 2911-14.

Uebel S, W.KRAAS, S.Kienle, K.H.Wiesmuller ,G.Jung, and R.Tampe.(1997) recognition principle of the TAP transporter disclosed by combinatorial peptide libraries. *Proc Natl. Acad.Sci.USA*; 94:8976-8891.

Warren EH, Otterud BE, Linterman RW, et al (2002) Feasibility of using genetic linkage analysis to identify the genes encoding T cell-defined minor histocompatibility antigens.*Tissue Antigens; 59*: 293-303

Warren D. Shlomchik.(2007) Antigen presentation in GvHD in MHC-matched allogenic stem cell transplantation . *Nature Reviews immunology; 7:340-352*.

Welniak LA, Blazar BR, Murphy WJ. (2007) Immunobiology of allogeneic hematopoietic stem cell transplantation. *Annu Rev Immunol*; 25: 139-70

Xin Feng, kwok min hui ,hasehem m,younes and Anthony G,Brickner. (2008) Targeting minor histocompatibility antigens in graft versus tumor or graft versus leukemia responses. *Trends in immunology*; 29(12):624-630.

yes **I want** morebooks!

Buy your books fast and straightforward online - at one of world's fastest growing online book stores! Environmentally sound due to Print-on-Demand technologies.

Buy your books online at
www.morebooks.shop

Kaufen Sie Ihre Bücher schnell und unkompliziert online – auf einer der am schnellsten wachsenden Buchhandelsplattformen weltweit! Dank Print-On-Demand umwelt- und ressourcenschonend produzi ert.

Bücher schneller online kaufen
www.morebooks.shop

Printed by Books on Demand GmbH, Norderstedt / Germany